The Fairburn Agate
of the Black Hills

100 UNIQUE STORIED AGATES

by James Magnuson
Photography by Carol Wood

Adventure Publications, Inc.
Cambridge, Minnesota

Dedication

To Betty and Tom Woodden

Edited by Brett Ortler
Cover and book designed by Jonathan Norberg

10 9 8 7 6 5 4 3 2 1
Copyright 2012 James Magnuson and Carol Wood
Published by Adventure Publications, Inc.
820 Cleveland Street South
Cambridge, MN 55008
1-800-678-7006
www.adventurepublications.net
All rights reserved
Printed in China
ISBN: 978-1-59193-294-9

The Fairburn Agate of the Black Hills

100 UNIQUE STORIED AGATES

Introduction

In this book you will see breathtaking images that feature the beautifully complex pattern sequences and color variations found only in the Fairburn Agate. Take your time and absorb the images of these stones; each agate has been painstakingly hunted and collected and is shown in natural light. Like all agates, Fairburn Agates are "lovers of light," and we have endeavored to place them in surroundings that let the stones speak to your appreciation of natural beauty. You will also find stories that encourage you to listen to the agates—perhaps they will tell you to go in a new direction or to go out on your own hunt. Fairburn Agates are some of the rarest agates in the world, and it takes patience, diligence and good fortune to find them. Because these gemstones are so beautiful and sought after, Fairburn hunters have many tales to tell. The stories about the agates in this book are told by their owners, Betty and Tom Woodden, and will give you a chance to walk in the footsteps of those who have sought and gathered these remarkable gems.

There are also several stories that provide visual vignettes of the diverse locales for Fairburn Agate hunting. Take note that for all of these locales, agate hunters must do their homework to determine what land is publicly accessible and what land is private and requires direct permission from private owners. You can obtain maps that outline the Buffalo Gap National Grassland (BGNG)—these public lands are generally accessible, but be sure to get a detailed map. In each of these areas you will find many other beautiful types of agates, petrified wood, fossilized coral and other fossils.

This book is only educational in the sense that you will see what natural forces are capable of producing. In this introduction, we will provide a simple and non-scientific discussion of the natural processes that resulted in the formation of these gemstones.

In the book's appendix, we provide information about the most sought-after characteristics in a Fairburn Agate. We provide this information to whet your appetite for personal exploration or simply to help you better appreciate the beauty of banded Fairburn Agates.

Most Fairburn Agates are a common type of agate known as a "fortification agate." Fortification agates exhibit concentric bands that resemble walled fortifications. In terms of chemical composition, Fairburn Agates consist of chalcedony, a silica-based mineral, and feature alternating bands of color. These color variations are likely caused by various mineral impurities in the chalcedony, such as iron or copper. Like all agate varieties from around the world, Fairburn Agates formed over a long period of time and there is a significant degree of speculation over how they formed. While we don't know the whole story yet, we do know that Fairburn Agates formed within sedimentary limestone, since many agates are found within this "host" material.

Over millions of years, cavities formed within the limestone, and silica gels that may have resulted from meteoric waters (rain) or the decomposition of marine organisms gradually seeped into the cavities, forming concentric bands. The hardened agate nodules were exposed by the Black Hills uplift, and over millions of years they were tumbled in riverbeds and streambeds, eventually eroding out of the softer host limestone. It is also speculated that Fairburn Agates were then buried in ash resulting from volcanic activity in the western United States and then finally weathered out again during the formation of the South Dakota Badlands.

Because of this unique formation process, Fairburn agates are as rare as they are beautiful. We hope you enjoy them as much as we do!

Cheyenne

LIKE MANY AGATES of stunning size, color, and intensity of banding patterns, this Fairburn Agate carries a name—"Mr. Fairburn." I went out to hunt for agates with a couple of my good friends on a Saturday in the Fairburn beds and found some nice agates. When you experience good fortune, you often feel a renewed sense of energy. Since we had made some good finds, we wanted to go back out on Sunday and we talked about hunting out along the Cheyenne River. Then my wife reminded me that we had ushering duties at church and I was unable to go. One of my friends decided to go out on his own anyway. I have noticed that in life there often seems to be a ripple or domino effect that stems from the things we choose to do. One choice leads to one outcome, while another leads to something different. Well, while we were at church performing our service, my friend found Mr. Fairburn on a gravel bar close to where we had hunted on Saturday. Maybe as a result of making the decision to forgo my desire to go out agate hunting, God found a way to reward us with this unique specimen—in this case a very positive "ripple effect." I now own this priceless treasure and truly believe God wanted us to find and share his beautiful creation.

Mr. Fairburn

Rolling Stone

Synchronicity

LOOKING BACK on over 40 years of Fairburn Agate hunting, I still remember the rock that got me hooked. It was in September, 1962, on the last Saturday of the month. I had just started working a job with the state in the engineering department, and was assigned to a survey crew. The party chief on the crew was named George. I talked with him over lunch and he said he was going agate hunting on the weekend and asked if I wanted to go along. I was intrigued and agreed to give it a try. We drove out east of Oelrichs and arrived at an area with agate beds. George went to the back of his truck, got me an army pick, and explained that I might need to work on the outer edges where the short grass is, because the easy finds have already been taken. He explained that if I noticed unusual looking rocks, or any kind of pattern in brown matrix, I should inspect it. After 20 minutes or so, I turned over a rock and it had a three-inch-long pattern. I called out to George and he came over thinking it would be a Prairie Agate, a very common and plain type of agate. Instead, he looked at it and said, "Wow, you son of a gun!" It was the best rock of the day, and beginner's luck for sure. That probably was the most important agate I've ever found. I refer to it as my "fishhook rock" because it got me hooked on hunting Fairburns. And I will always be grateful to George because he's the one who introduced me to this wonderful hobby.

Hook, Lines and Sinker

1880

Alpine Slide

ROCK HOUNDS spend a lot of time outdoors and develop a comfort level with all things natural. A story that illustrates this nicely involves an occasion when my wife's sister Deanne came hunting with us. After a while she found a rock, brought it over to my wife for inspection, and asked whether it was a Fairburn Agate. My wife took the rock, looked it over, then she licked it and took a closer look. Her sister was a little surprised and asked why she was licking a dirty old rock. My wife simply said, "That's what you do."
I don't think Deanne really believed her, so without telling me about what had happened, they gave me the rock and sure enough, I licked it! They both started to laugh and I didn't know why. After they filled me in I had a good chuckle as well, and yes, it was a nice little Fairburn.

Painted Turtle

Ginger Blossom

Old Flame

DURING THE EARLY '60s, while living in Hot Springs on the south end of the Black Hills, I often went hunting for "road agates." I had been given my grandmother's 1951 Buick and used it as a second car. I needed to occasionally start the car up and run it a bit, so I'd take it for a drive in south Fall River County and hunt the gravel roads south and west of the town of Oelrichs. At the local gravel pit the processing included a screening unit that sorted the stones by size and saved them from being crushed. It also preserved many nice agate specimens. This gravel was used along the secondary roads and the larger rocks always found their way to the shoulder of the road. I have many nice Fairburns in my collection that were found on my jaunts in the old Buick.

Silver Lining

Sea Level

Salt and Pepper

I HAVE A TWIN BROTHER (Tim) who lives quite a long ways away but comes home to South Dakota for occasional visits. When he visits, he enjoys getting outdoors and is always willing to accompany me on a Fairburn hunt. I can remember the first time I ever took him with me. It was the 1970s, and we went to one of my favorite spots near the Fairburn beds. We were working some hilltops and had been hard at it for almost three hours. My brother was about 200 yards away and hollered out to me, "Hey, I think I found one!" I gingerly but quickly worked my way over to where he was and he was holding a two-inch beauty. There's nothing better than being out with friends and family on the occasional hunt and seeing them experience the thrill of finding an awesome Fairburn.

Outback

Fabergé

Daytona

I HAVE BEEN FORTUNATE to be schooled by old-time Fairburn hunters. Experience in knowing when and where to hunt are two of the critical success factors for finding gemstones that are as rare as these agates. My seasoned mentors also taught me to be able to spot some of the "companion stones" that indicate that you are looking in gravel deposits that may also contain Fairburn Agates. In particular, you need to be on the watch for naturally tumbled pink and white quartz, and, if you see a significant number of these stones, it's likely that you're in the right area to find Fairburns.

Peace in the Valley

Cuts Like a Knife

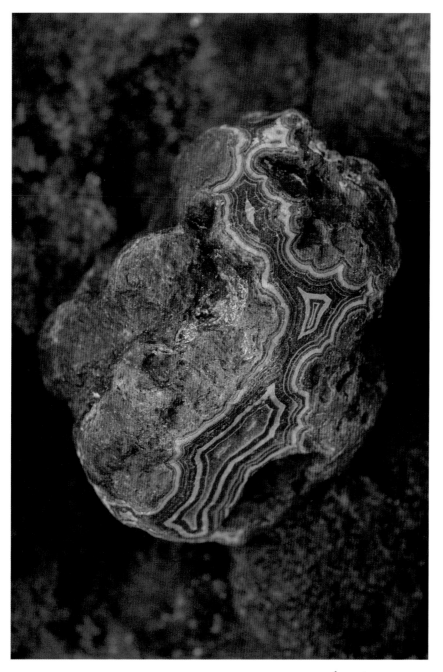

Born in the USA

THE FAIRBURN, OELRICHS, AND REDSHIRT CREEK hunting areas begin about 30 miles south and slightly east of Rapid City—running farther south and east another 15-25 miles. These hunting locales share common features, including a hilly, rugged terrain that is dotted with "bald hills" that contain weathered-out gravel deposits. With little or no vegetation, these hilltops can resemble a barren moonscape, but they are excellent for Fairburn Agate hunting. The vegetation surrounding these hills is a mixture of prairie grasses and cactus; be sure to wear heavy-duty boots for both climbing and hiking. There are also numerous sinkholes in this area—some of which are quite large—so be vigilant as you drive into these areas. The topsoil consists of light-colored limestone sand and sediment that is commonly referred to as "gumbo," because it rapidly transforms from a hard, crusty material to a thick, pasty "gumbo" when it rains. Many unprepared novice agate hunters have found themselves stranded when their vehicles became mired in this mud.

Purple Haze

Key Lime

Vesuvius

ON AN EASTER SUNDAY in the late '60s, we planned to go out for a family Fairburn Agate hunt after church. We even got my mom (Clara) to go with us, and it was the first time she came along on the hunt. After enjoying a picnic dinner, we packed things up, spread out, and started to hunt. After a couple hours we all met back at the vehicle and no one had found much. We had all covered quite a bit of territory—except for my mom, who had stayed close by the vehicles and slowly covered the gravel deposits in a shallow ravine. As I was returning, mom came up to me and said, "Tommy, is this one?" It was a small but beautiful Fairburn, and certainly the best of the day. I can't say that mom became a regular on my hunting expeditions, but I can say that she had the eye for finding these earthly treasures.

Tomahawk

Hickok

Sunspot

DURING MANY PAST WINTERS, my wife's father (Les) ran a trapline for beavers and muskrats in the Hat Creek area. I often went with him to open gates and help out as needed. In the vicinity where he did his trapping, there are gravel bars that drain from Nebraska toward South Dakota. These bars aren't very large and are often covered with snow during the winter. On one outing I stopped by a gravel bar while waiting for him to check the trapline and, while intermittently hunting and striving to stay warm, I found a nice brown and white agate, which warmed me up nicely (at least on the inside).

Great White

Over Easy

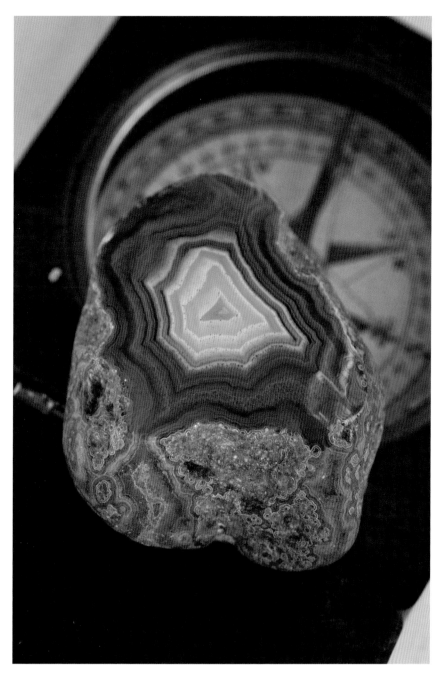

High Plains Drifter

FAIRBURN AGATE HUNTING is not without its risks. In recent years some work has been done to dig up snake dens out in the grasslands in an effort to root out the more aggressive serpents. The area is home to the venomous prairie rattlesnake and is also habitat for the non-venomous bullsnake, which is often confused with the rattlesnake because it too makes a rattle-like sound when threatened. Some people have speculated that bullsnakes (which tend to be more aggressive) have interbred with rattlesnakes, but professional zoologists have dispelled this myth. Rattlesnakes are live bearers while bullsnakes lay eggs, and the two snakes are biologically incompatible. A well-advised hunter will wear snake gators—hard rubber boots that go almost to hip level. Some hunters will even bring a sidearm. However, the best way to avoid rattlesnakes is to simply stay alert when walking through the grasslands, when stooping to pick up a stone, or when there are snake holes in the vicinity. (And if you see a snake, change direction, of course.) The second most critical thing is to know what to do in the unfortunate event that you're actually bitten. With this said—venomous snake bites are even more rare than lightning strikes on the golf course and certainly more rare than Fairburn Agates!

Rattler

Scene of the Crime

Chief

THE LAME JOHNNY CREEK hunting range begins about 15 miles south of Fairburn along Highway 79. The creek flows out of the southeastern end of Custer State Park and forms a tree-lined corridor surrounded by native prairie grasses on rocky slopes and hilltops. With the Black Hills as a nearby backdrop, this is a scenic area to casually walk the shallow streambed while hunting for Fairburn agates. Note that the streambed is dry for most of the year, and it is best to hunt in the spring after there has been snowmelt and spring rains that bring new stones into the creek, and turnover long-hidden faces of beautiful Fairburn Agates. Your patience will be rewarded if you take the time to slowly walk and peruse these ready-made gravel beds. When your feet grow tired, have a seat and turn some stones over—you just might hit the jackpot! Please also note that much of the land that the Lame Johnny creekbed traverses is private, so double-check before setting out to hunt.

Longhorn

Draw

Ring Around the Rosie

A LOCAL APARTMENT building manager was getting complaints from tenants about a leaky roof and decided to remove all of the roofing gravel in order to begin the repair. A young boy was drawn to the gravel pile and began to climb around on it. He spotted this full-patterned three-inch Fairburn and knew that he had found a real treasure. As I sometimes do, I made a fair deal with him and I am now proud to display this agate among my best finds. I am happy that it was rescued from a fate of sitting among countless other featureless stones atop a building complex, and I know you will enjoy gazing at this stone and be thankful for the young man who decided to climb around on a dirty old rock pile.

Pilgrim

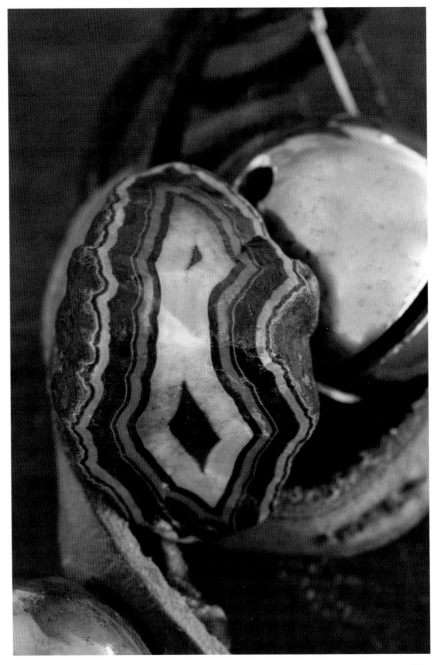

Bandit

Grass Roots

MANY EXPERIENCED agate hunters will tell you that their success is proportional to their level of effort and diligence. However, they will also tell you that there is an element of luck and it's impossible to predict where (or how many) Fairburns you'll find. Despite this, the very same people will tell you that if you go through their collection, a disproportionate number of their good finds were found within the first 30 minutes of the hunting day. One such find was made when one my buddies and I pulled up to a gate between two pastures where we were planning to hunt. After hopping out of the vehicle to open the gate, I looked down at the ground and there lay a beautiful agate the size of a fifty-cent piece. I almost felt a little guilty at having snagged this beautiful gem with so little effort, but then I quickly remembered countless hours of exhaustive hunting with no luck at all and I happily pocketed my early-day find.

Odyssey

Oh Christmas Tree

Bermuda

ONE OF MY American Indian friends once brought me a stone that I could tell—by looking at the husk—was likely a Fairburn agate. It was almost black and there was a small hole going into the rock. He brought the agate to me because he wanted me to cut it in half and see if there was any pattern on the inside, so he left it with me. When I got around to cutting it open, I couldn't have been more stunned by the rich and vibrant colors awaiting me. I gave my friend a call and at first I gave him some good-natured ribbing by telling him that it turned out to not be much of a rock, but he might as well come pick it up. Needless to say, he was just as stunned as I was and took the agate to share with his friends. After a while he brought it back for me to face polish, which enhanced its beauty even more. I told him that if he ever wanted to sell the stone, he had to sell it to me, and I'm happy that I was eventually able to purchase this beautiful specimen.

Monarchy

Shetland

Neapolitan

EXPERIENCED FAIRBURN agate hunters will say that gravel pit hunting near the Cheyenne River is highly productive. In days gone by, the pits were a welcoming place for agate enthusiasts. Now the ownership has passed on to the next generation and times have changed. Agate hunters are no longer welcome. Signs are posted that read "No Trespassing" or "No Rock Hunting" and pits are patrolled by the sheriff's office. When one asks, the common refrain is that collecting is prohibited because of insurance issues and acts of vandalism. Whatever the reason, it has forced agate hunters to be ever more resourceful and persistent. Just remember—all that gravel ends up somewhere!

Gold Rush

Old English

Cardiac

MANY PEOPLE who are new to agate hunting search randomly and therefore hunt ineffectively on gravel piles, sand bars and rock-strewn hillsides. They will walk about aimlessly and in a very short period of time come to the conclusion that there are no agates to be found and move on to the next gravel bar or pile. Over many years of hunting, I have learned that it's critical to work a gravel deposit systematically and keep the sunlight at your back as much as possible. In fact, I've learned to slowly walk in a grid pattern that gives maximum coverage. As Fairburn Agates become more sought after, and therefore more rare, one needs to look for smaller and smaller clues in terms of color, pattern and host stone characteristics. At times, this can seem laborious, but when it yields results, I'm quickly reminded that it's a labor of love.

Labyrinth

Root Beer Float

Kerchief

FAIRBURN AGATES are often found still embedded in the host sedimentary rock they formed in. Many other agate types, such as Lake Superior Agates, are rarely found this way and often have completely weathered out of the host material. Because Fairburns are often still surrounded by this other material, if the exposed pattern is face down, you'll never know it's there, as you'll simply see a tan, brown, black or gray chunk of weathered stone. On days when I've become exhausted walking across gravel-strewn hills or river bars, I'll get down on my knees and patiently turn over a lot of rocks. This is the ultimate test of the patient agate hunter and once in a while it leads to success. When you think about how long it took for these gemstones to form, weather out of their original location, and be naturally tumbled smooth, the time and effort spent patiently seeking them is insignificant indeed.

Husker

Red Skies at Night

Rushmore

THERE ARE AGATE hunting grounds near the small town of Interior, which is about 60 miles southeast of Rapid City—and located near Badlands National Park. This is rugged and generally arid terrain, so take care to maintain your footing when climbing up or down hillsides. There are also more flattened-out areas with large deposits of naturally tumbled gravels that yield Fairburn Agates. As you travel through the area and agate hunt, you will be treated to unearthly beautiful scenery that includes sawtooth ridges, badland hills and canyons, gorges, and towering white limestone spires. You will also find Petrified Wood and Bubblegum Agates to be abundant in this hunting locale, and this area is notable for many kinds of ancient marine fossils.

Polar

Follow the Yellow Brick Road

Cashmere

IN ONE of my stories, I mention that a large number of my best finds come within the first half hour or so of hunting. It's equally true that a large number come within the last half hour of the day. Perhaps our senses are sharpened when we've been anticipating the hunt (early day finds), or when we know that we only have a short time left to make that memorable find. On one outing I pushed the limits a little bit and was still hunting after the sun had set. It was almost dark and I decided to make a final pass across a 200-foot-long gravel bed. As I was straining my eyes to pick up any Fairburn Agate characteristics, I spotted some exposed pattern and picked up a beautiful gem near the end of the exposed gravel. When I got back to my vehicle, I was able to get a little more light on the agate and it was an excellent specimen that easily justified hunting in the fading light.

Wanted

American Girl

Tumbleweed

WHEN I WAS working full time, my job required a fair amount of travel. When I knew that we'd be going through areas with the kinds of gravels that might contain Fairburn Agates, I'd usually pack a lunch should I have an opportunity to do some rock hounding during my break time. On one such day I stopped by an old dry creekbed that comes down from Teepee Canyon (famous for another beautiful variety of Black Hills agate). I spotted a nice gravel bar a short ways off and walked over there. I hadn't been hunting for 15 minutes when I spotted a two-inch prize gemstone. When I think back on that experience, the old saying of "The worst day of agate hunting (or fishing, golfing, etc.) is better than the best day at work." On this day I decided the saying should be "The best day at work is when I get to mix in some agate hunting and find a quality Fairburn."

A Window to Your Past

Harvest Moon

You-Tube

A LARGE PORTION of the land where Fairburn Agates are found is within the boundaries of the Pine Ridge Indian Reservation. I had a good friend who was a Lakota tribal member from the reservation, and he was an avid agate hunter. He sometimes attended shows hosted by our local rock club, which was formerly known as the Picture City Gem & Mineral Club. At our shows we always liked to display our new finds and talk about where we'd been hunting. I remember one conversation with my Lakota friend and told him there were times when I was out on the gravel-strewn hills or river bars and I didn't know which way to turn. He said sometimes you need to be very quiet and just listen, and the agates will talk to you and tell you which way to go. He said that at times he has heard a soft rustling, walked in the direction of the sound, and found it would lead him to a Fairburn Agate. This hunting experience teaches us to slow down and pay closer attention to our surroundings.

Wowahwa

Patchwork Quilt

Old Blue Eye

WHEN HUNTING FOR Fairburns in the grasslands and badlands, most of the underlying surface material is dried and hardened alluvial material (i.e., sedimentary sands and silts). For the inexperienced or occasional visitor to this area, they are well advised to get a current weather forecast before heading out on the hunt, and to keep a close eye on the sky for signs of rain. Storms can come about quickly in southwestern South Dakota, and if you are still out in the grasslands by the time it starts to rain, you may well be stranded for several days. The rainwater quickly transforms the surface material into a heavy paste that the locals refer to as "gumbo," and even a sturdy 4x4 vehicle will get bogged down. At that point you can either hike out (usually a considerable distance) or wait for the ground to dry (usually a couple of days if no more rain comes). If you have provisions and can contact loved ones by phone, the latter option may just be the perfect opportunity to spend some "quality time" hunting among recently washed gravel deposits!

Candy Striper

Opening Line

Serape

THE RAILROAD BUTTES are located in the heart of the Buffalo Gap National Grassland (BGNG); as mentioned in the introduction, you can readily acquire maps for the public lands that are open for hiking and agate collecting. This hunting range begins about 20 miles southeast of Rapid City and extends farther south and east for another 20 miles towards the town of Scenic. This area is horse country and as you drive the narrow and winding gravel roads you will surely see horses and ponies grazing and galloping across the scenic hills and valleys. Nestled among the table-flat buttes and grassland you will find highly eroded hilltops and flats with vast quantities of naturally tumbled stones that weathered out of the Black Hills. This area is easily accessible and you can walk for miles in any direction, but take care to navigate by compass or GPS, identify landmarks, or keep a sight line on your vehicle because the vast expanses of similar terrain can lull you into directionless wandering. Also, take care to be selective in collecting the large and beautiful Prairie Agates that are abundant in this area or you'll find yourself slowly unloading your (increasingly heavy) treasures as you trek back to your vehicle.

And Then There Was Light

Day's Limit

Masquerade

LIKE MOST SERIOUS HOBBIES, agate hunting has a competitive element. During the years when Fairburn Agates could be found more readily, I would go out hunting with my friends. To increase motivation, we'd have a contest to see who could find the most agates in one day. The rules were simple: for the agate to be counted, it must have at least ¾ of an inch of pattern. From the late 1960s through the early 1980s, we could sometimes find 10-15 Fairburns in one day! The most I ever found in one day (of this size) was 11; I was hunting the gravel bars along the Cheyenne River. Those are some great memories and I sincerely hope hunters will have this kind of success again in the future.

Caribbean

Cumulonimbus

Little Bighorn

WHEN HUNTING for rare gemstones, one spends a lot of time with their own thoughts, and this is especially true when you're not having much luck. On one such day I was out with a couple of friends hunting along the Cheyenne River gravel bars (a great place to be even when you're not having any luck). My friends were off in the distance wandering somewhat aimlessly while I systematically worked successive sections of gravel. As I sometimes do, I silently said, "Lord, give me a little help," and then I stopped to listen as my Lakota friend had taught me. I became more relaxed and perhaps less intense in my concentration. Within a short time I spotted a very small surface feature and retrieved one of my best-ever Fairburn Agate finds. I don't know if it was my silent prayer, my listening, my systematic searching method, or just dogged persistence, but I do know that finding these gems is a life passion that I'll always enjoy.

Dream Catch

Mahogany

Rock Star

THE TERRAIN in the badlands and grasslands is highly diverse, as is the corresponding mineralogy. A couple of years ago I was out hunting with a few friends on the high ground east of Buffalo Gap and along the Cheyenne River. It was one of those days when I hadn't found anything of note despite my diligence and systematic approach to hunting. I decided that it was time to change my hunting area and started hiking. I cut across a draw, through a sparse grove of cedar trees along some shale banks. I looked down at the ground and there lay a beautiful agate all by itself amid the shale. It had likely rolled down a hill and come to rest, and there it sat just waiting for me to find. I left the agate sitting where I spotted it and called my friends over to witness my good fortune of finding a fully exposed Fairburn by itself on the shale.

Primary Colors

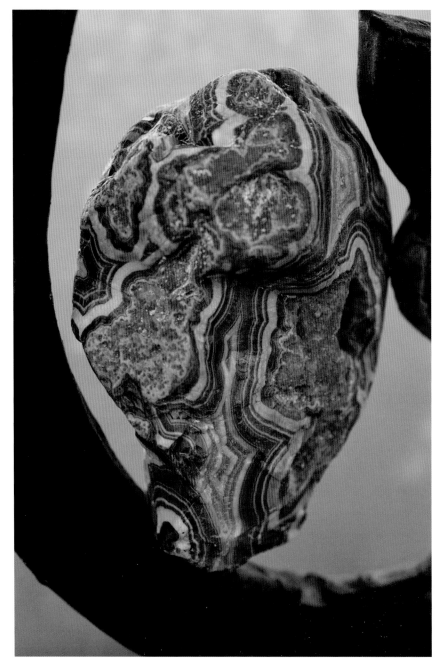

Scarlett

High Roller

THE HUNTING RANGE near Indian Creek is within the Buffalo Gap National Grassland (BGNG), and begins about 5 miles south of Creston. The creek runs north and west for about 10 miles through the grasslands before emptying into the Cheyenne River. This area is distinguished by more distinctive and sharply sculpted bluffs and buttes than in other hunting grounds. The area features stunning panoramas that are reminiscent of the desert Southwest. The shallow creekbed (which is often completely dry) meanders lazily across the flat landscape below the bluffs. Beneath an often cloudless sky, you can walk this creekbed for miles. In addition to Fairburn Agates, you may find some nice black-and-gray Prairie Agates here that display more detail than their caramel-and-white cousins. This area is also home to the Black Agate, which has jet black chalcedony highlighted with bright white banding features. Bagging a few of these choice Prairie and Black Agates will keep you going as you wander along the streambed and hunt in the abundant gravel deposits.

Bat Cave

Mustang

Born to Be Wild

ONE OF MY longtime Fairburn hunting friends (Elmer), also a dear friend of our family, went out hunting with me in the Redshirt Creek area (which was formerly known as Schumacher Creek). As we were hunting, a thunderstorm came up quickly, and we were about three miles from a gravel road, so we had to hustle out since we were parked on ground that would quickly turn to gumbo mud if it started to rain in earnest. After reaching the truck, Elmer said there was a gravel road we could park on nearby that would stay solid. By the time we reached our destination, it had started to rain pretty hard. We hopped out and started walking through the gumbo mud and our feet got caked with the heavy, sticky paste until they were about twice their normal size! We separated and started to hunt, knowing that the rain-moistened rocks provided a great hunting advantage. After a short time I heard a rock hammer striking stone and then "the air turned blue." As I worked my way over to Elmer, I could tell it would not be good. Sure enough, he was picking up fragments of what had been a beautiful Fairburn Agate. Elmer recounted that the agate had been covered with a lot of mud and he thought it was probably just a piece of red jasper, so he decided to break it open. While we found a couple of other nice agates, none were as good as the shattered Fairburn. The moral of the story: be careful how you use your hammer.

Dances with Color

Luck of the Irish

Back in Black

AS YOU READ through my stories, you will note that I have been blessed with family and friends who have been willing to share in my passion for hunting Fairburn Agates and other marvelous gems and fossils from South Dakota. The vast majority of my family members' finds (and some of my friends' great finds) are now in my collection. I think they have entrusted me with these beautiful gems (either by giving them to me or allowing me to purchase them) because they know that they will be kept together, cared for, and available for display as long as I am here on earth. These colorful creations are part of me and each one is a memory of the times I spent with agate hunting companions.

National Treasure

Ghost

Poplar Guy

MY WIFE BETTY and I have enjoyed some quality time together hunting for agates, but one occasion stands out above all others. We were hunting in the Redshirt Creek area, and my wife Betty found one half of a nice Fairburn that had been broken relatively recently, possibly when weathered out of a deformation channel, or perhaps Native Americans had broken the stone trying to make some type of ornament or implement. She hollered at me to come see it and it was a nice specimen that was about two to three inches in diameter. We went back to hunting and about 30 minutes later I found the other half about 40 feet away. We were in an area of rolling grass hills with knobs on top, and the grass was relatively thick, so finding these companion agate halves was truly remarkable. This agate has come to be known in our home as "half hers—half mine."

You Complete Me

Tuscany

Upstream

THE CHEYENNE RIVER cuts a neat path straight through the heart of the Fairburn Agate hunting range. In this area, it runs on a line from Edgemont on the southwest end of the range to Wasta at the north end. The Cheyenne River Basin is perhaps the most verdant terrain that you'll find in the Fairburn range, with more trees and native prairie grasses than in other areas. Because of the large expanse of land that the river traverses, you will also see a broader range of geological features, including some beautiful panoramas of steeper cliffs and buttes. Along the Cheyenne, there are numerous gravel mining operations where you'll be tempted to stop and hunt for agates, but you must request permission from owners or operators before entering these operations. There are also numerous sand and gravel bars that are continuously being covered or uncovered as the course of the river shifts. If you are able to get a kayak or canoe, you can have access to some of the best Fairburn Agate hunting in the area and will certainly be hunting in places that are less accessible to those hunting on foot.

Jailhouse Rock

Buddy Holly Leaf

Río Grande

I MENTION in one of my stories that hunting for agates has a competitive element. This agate was found while I was hunting with a friend and, as was often the case, he was getting out ahead of me while I systematically worked over areas that he had already walked through. I eventually spotted this beautiful Fairburn Agate and shouted ahead to my friend, "Hey, aren't you going to pick this one up?!" Of course he took it as good-natured ribbing—and as is true with agate hunting buddies—we'd always rather have one of us find one of these stones than to have never found it at all.

Sundance

Look Into My Crystal Ball

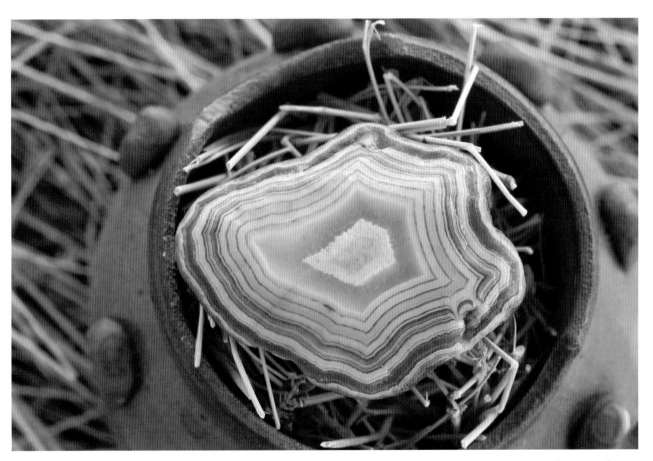

Eye of the Tiger

PERHAPS THE MOST important story to tell is about coming up empty and not finding a Fairburn Agate. It's a universal experience for those that hunt these beautiful gems. In fact, finding a Fairburn is sometimes likened to finding a red-colored grain in a bucket of sand. Many serious agate collectors have been successful in hunting for rare and elusive rocks, fossils and minerals and have learned to be a student of each new type of specimen. One improves their chances by taking the time to find actual stones (in stores or in personal collections) or quality photos of rough stones, and by researching hunting locales and protocols. Meeting with local hunters and collectors who are willing to share their time and insights is often especially helpful. There are numerous veteran agate hunters in western South Dakota who will warmly welcome you into their homes to share their finds, their stories, and the tricks of the trade for hunting Fairburns. But even with this thoughtful preparation, be prepared to trek across countless miles of grasslands, along dry creekbeds, in active gravel pits, on sand and gravel bars of rivers and streams, and still come up empty-handed. At times one can become dispirited and even claim to be "cured" of the agate hunting bug, but then they see a beautiful Fairburn like this one, and realize that they will come back, and they will hunt again until they find their own prized Fairburns.

Come Together

Table Rock

I Will Always Be Your Friend

Appendix

Fairburn Agates exhibit a stunning array of color combinations as well as intricate holly leaf-like patterning, a combination that makes them both beautiful and valuable. While there are many specific types of other well-known agates (such as the Lake Superior Agate), there aren't as many varieties of Fairburns. However, there are other unique aspects of these rare gemstones that can help the collector distinguish them and enhance our appreciation of their natural beauty.

One aspect to consider is how to assign monetary value. World-class gems, such as those shown in this book, command high market prices primarily because of the elements shown below:

Patterning: Fairburn agates that display either a full or complete face and have a pattern that "wraps around the entire stone" have a higher value than those that are partially surrounded by host limestone/matrix.

Color: Agates with the deepest colors are most valuable; stones exhibiting bright orange, deep red or pink, or jet black are especially sought after. This is also true for stones with multiple color variations; there are some Fairburn specimens with up to a dozen different unique colors!

Intricacy: The "holly leaf" pattern of Fairburn Agates is what makes these stones distinctive. Specimens with the most intricate lines and patterns, especially those with deep offsetting colors, are most valued.

Banding: Fairburns with tight and more-defined bands provide greater visual depth and intensity, making them more sought after.

Another unique characteristic of Fairburn Agates is how they obtained their color sequences. Like other types of agates, their surface coloration is the result of two things—mineral impurities that mixed with the silica-based liquids from which they formed, and the materials they were exposed to after weathering from their host limestone. Specifically, Fairburn Agates were often "reinterred" beneath Black Hills alluvium, including volcanic ash. Mineral impurities occurring both during and after formation include copper (lending shades of blue or green), magnesium (yielding the deep black colors), and iron (providing numerous reddish, orange and pink hues). While the beauty of the Fairburn Agate is certainly more than skin deep, it is best to leave them in their natural state rather than attempt to polish them, as some of the more robust and distinctive color combinations may well be those at surface level.

Lastly, it seems that some Fairburn hunting locations tend to yield agates with a generalized set of the characteristics and colors described above. This may be due to unique weathering conditions or mineral characteristics of the soils they became embedded in, or possibly the areas within the Black Hills that they originally eroded out of. The photos below show some Fairburn Agates and the hunting locations where they were found. While these agates are characteristic of many specimens found in these localities, there is no hard-and-fast guide to what can be found in a given location, especially with continued erosion and movement of the stones along rivers and creekbeds. Below you can see some examples of Fairburn Agates from specific localities.

FAIRBURN

CHEYENNE RIVER

LAME JOHNNY

INTERIOR

NEBRASKA-CRAWFORD

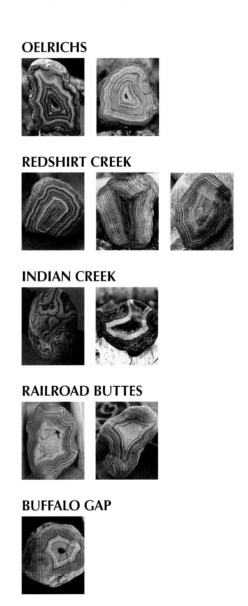

OELRICHS

REDSHIRT CREEK

INDIAN CREEK

RAILROAD BUTTES

BUFFALO GAP

About Contributors Tom and Betty Woodden

The heart and soul of this book belongs to Tom Woodden and his wife Betty; their personal passion for Fairburn Agates and other treasures of western South Dakota are on display both in the photographs and in the stories in this book. How does one develop a lifelong desire and joy of hunting and collecting things from the earth? It might be sparked by another person or by an experience from one's youth or from something within. Tom will tell you that many people who enjoy hunting for agates share the trait of quiet calmness, which allows them to appreciate nature at a deeper level. And their personal humility and a belief in God allow inspirations to come from Him through natural wonders.

As Tom grew to enjoy and collect beautiful Fairburn Agates and other minerals and fossils, his relatives and friends supported him and added to his collection. Tom's mother (Clara) often spent time with him in his workshop. She had a true love of Fairburn agates, and she loved to watch him work the agates through cutting, polishing, and fashioning into jewelry. Most of all, though, "Mom" loved to just pick up the agates, look at them with a magnifying glass, and often remark that she thought the agates truly were a work of God.

Betty and Tom are not only gemstone and fossil collectors, they are ambassadors for rock hounding. They frequently show and sell Fairburn Agate specimens and other gemstones and fossils at locales across the U.S. and internationally, including at a major show in Germany. Tom and his friends also organize the premier western South Dakota Fairburn Agate Show—the annual Outdoor Rock Swap in May—which brings out Fairburn enthusiasts from around the country, and allows newbies to learn about agate hunting and gives them the opportunity to purchase agates for their own collections. Tom and Betty's collection is world-class and it is a family heirloom that reflects a shared appreciation of beautiful gemstones. Tom is most happy when he knows that others are able to enjoy the fruits of his labor. His spark is infectious and we hope these photographs and his colorful stories/tales will light your inner fire and perhaps encourage you to seek your own natural treasures.

About the Author and Photographer

Jim Magnuson

Rock hounding is more than a hobby for author Jim Magnuson, it's a serious and rewarding avocation that helps him connect with nature. He has been an avid hunter and student of various gems, minerals and fossils since his childhood when he first began to hunt for stones in his native state of Illinois. In addition, Jim enjoys sharing his passion not only through showing and gifting some of his finds, but also through writing, another life-long interest. While much of his writing and presenting is done in a professional setting, he has recently begun to develop his creative writing talents through storytelling that is centered around his rock hounding adventures. As Jim has begun to share his enjoyment for and knowledge of rocks and minerals, he has also become more engaged with rock and mineral clubs as a way to further his learning and branch out into other types of agates, gemstones and geology. Jim will gladly offer his insights in areas that he is knowledgeable and listen intently to others who have gone before him in different rock hounding domains. It is an endless journey that he hopes to continue sharing with those who find beauty and wonder in earth's treasures.

Carol Wood

Carol Wood took up professional photography as a means of satisfying a lifelong passion for creating and sharing things of beauty. She has a keen eye for seeing perspectives in things that on the surface appear to be mundane or quite simple. Given her training and natural instincts for complementary color and patterns, Carol has been able to take beautiful gemstones and turn them into "paintings" that are stunning to avid rock hounds and to those who have never had an appreciation for rocks and minerals. In addition to Carol's photographic pursuits, she also enjoys outdoor activities with her friends and family, especially activities that have both a mental and physical component. As a result, she has become an avid rock hound in her own right and has gradually built a collection of beautiful agates that adorn her home in northern Illinois. Being able to work on this book was a win-win as Carol has a great fondness for the Black Hills and found Betty and Tom Woodden to be the perfect partners.

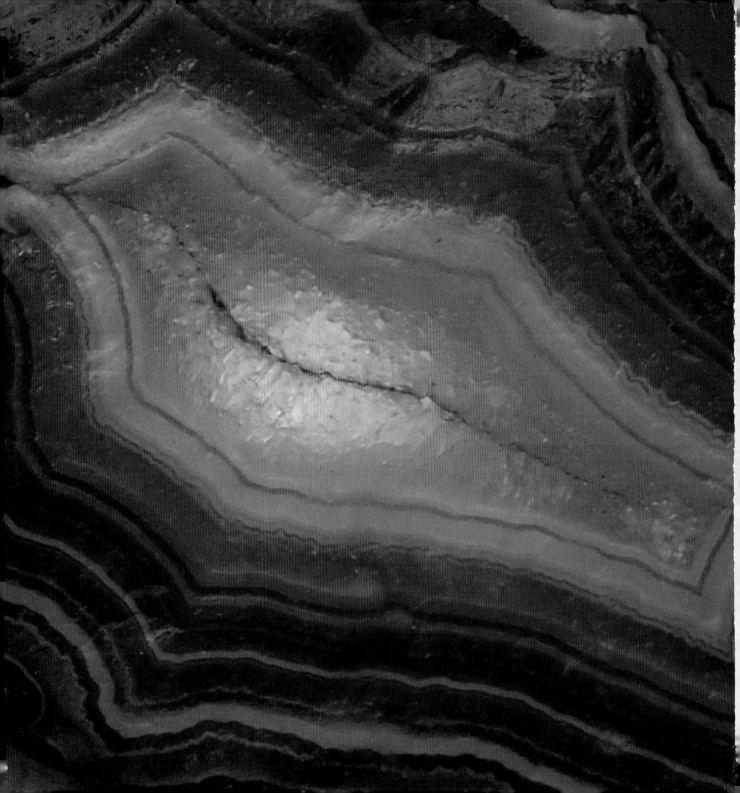